CRITICAL THINKING & LOGICAL REASONING WORKBOOK-9

9

GIFT OF LOGIC™ SERIES

An Essential Resource for Everyone

Boost Your Thinking Skills

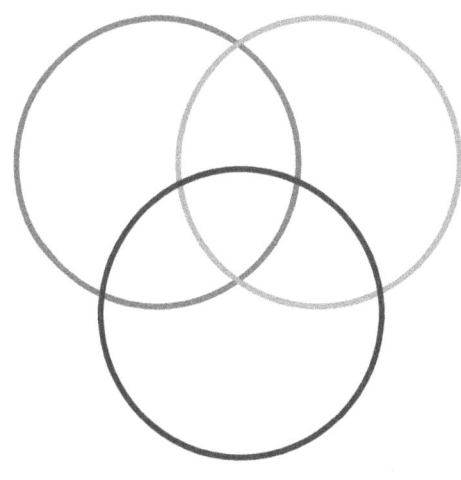

Verbal Reasoning

Analytical Reasoning

Pictorial Reasoning

THIRD EDITION

| FOR GRADES 6-12 | STUDENTS, TEACHERS, AND PARENTS |

Ranga Raghuram

GIFT OF LOGIC™

Gift Of Logic, Inc
http://www.giftoflogic.com
sales@giftoflogic.com

Critical Thinking and Logical Reasoning Workbook-9
ISBN-13: 978-1494833022
ISBN-10: 1494833026

Third Edition
1-2014

Copyright © 2009 Gift Of Logic, Inc. All rights reserved. No part of this publication may be reproduced, stored in a retrieval system, transmitted in any form or by any means, electronic, mechanical, photocopying, recording or otherwise, without the written permission of the publisher.

License: This book is licensed for use by one person only. Use of this book in a group setting (classroom, workshop, etc) without the written permission of the publisher is prohibited. Unauthorized duplication is strictly prohibited by law. Contact the publisher at sales@giftoflogic.com for classroom/school/group licensing.

GIFT OF LOGIC™
CRITICAL THINKING & LOGICAL REASONING CURRICULUM
12 WORKBOOKS TO BOOST YOUR THINKING SKILLS

For Kindergarten, Grade 1, and Grade 2

Workbook# 0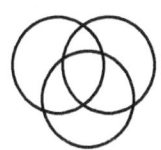

Verbal Reasoning	Finding the truth, Inferencing, Analogies, Synonyms and Antonyms, Agree/Disagree
Analytic Reasoning	Memory drill, Decision making, Positioning, Sudoku
Pictorial Reasoning	Connect the dots, Mazes, Picture Sequence, Spot the difference, etc

Workbook# 1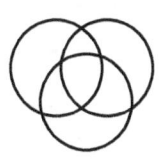

Verbal Reasoning	Finding the truth, Inferencing, Analogies, Synonyms and Antonyms, Agree/Disagree
Analytic Reasoning	Sorting, Positioning, Picking, Assorted problems, Numeric and Alphabetic Sudoku
Pictorial Reasoning	Picture Sequence, Spot the difference, Odd picture

Workbook# 2

Verbal Reasoning	Finding the truth, Classification, Direct and Inverse relationship, Inferencing, Analogies, Agree/Disagree
Analytic Reasoning	Sequencing, Scheduling, Strategy, Picking, etc
Pictorial Reasoning	Picture Analogy, Odd picture, Pattern matching, etc

For Grade 3, Grade 4, and Grade 5

Workbook# 3

Verbal Reasoning	Not, And, Or, If .. then, Conditional inferencing, Unconditional inferencing, Symbolic Logic
Analytic Reasoning	Lists, Sequencing, Grouping, Venn Diagrams, Graph logic, Number logic, Letter logic, Sudoku
Pictorial Reasoning	Picture sequence, Picture analogy, Odd picture, Picture difference, Pattern matching

Workbook# 4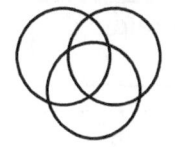

Verbal Reasoning	Contradiction, Converse, Inverse, Contrapositive, Conditional inferencing, Symbolic Logic
Analytic Reasoning	Scheduling, Looping, FIFO, LIFO, Correlation, Venn Diagram, Graph logic, Number logic, Sudoku, etc
Pictorial Reasoning	Picture sequence, Picture analogy, Odd picture, Picture difference, Pattern matching

Workbook# 5

Verbal Reasoning	Biconditional, Categorical inferencing, Cause and Effect, Symbolic Logic, Agree/Disagree, Word and Sentence analogy
Analytic Reasoning	Correlation, Grouping, Venn Diagrams, Graph logic, Number logic, Letter logic, Sudoku, etc
Pictorial Reasoning	Picture sequence, Picture analogy, Odd picture, Picture difference, Pattern matching

********* Essential resource for everyone *********
*http://www.giftoflogic.com *sales@giftoflogic.com

GIFT OF LOGIC™
CRITICAL THINKING & LOGICAL REASONING CURRICULUM
12 WORKBOOKS TO BOOST YOUR THINKING SKILLS

For Grades 6-12, College/University Students, Adults

Primer

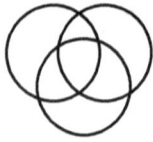

Prereq

Verbal Reasoning	Logical Operators, Conditional, Categorical and Causal reasoning, Validity, Fallacies, Symbolic Logic
Analytic Reasoning	Positioning, Grouping, Sudoku
Pictorial Reasoning	Pattern perception, Figure formation, Paper folding and cutting, Figure matrix, Rule detection

Workbook# 6

Verbal Reasoning	Arguments-Main point, Must be true, Cannot be true
Analytic Reasoning	Positioning, Grouping, Sudoku
Pictorial Reasoning	Pattern perception, Figure formation, Paper folding and cutting, Figure matrix, Rule detection

Workbook# 7

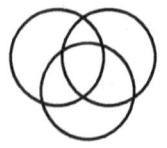

Verbal Reasoning	Arguments-Strengthening, Weakening
Analytic Reasoning	Positioning, Grouping, Sudoku
Pictorial Reasoning	Pattern perception, Figure formation, Paper folding and cutting, Figure matrix, Rule detection

Workbook# 8

Verbal Reasoning	Arguments - Controversy, Paradox
Analytic Reasoning	Positioning, Grouping, Sudoku
Pictorial Reasoning	Pattern perception, Figure formation, Paper folding and cutting, Figure matrix, Rule detection

Workbook# 9

Verbal Reasoning	Arguments- Assumptions, Reasoning strategy
Analytic Reasoning	Positioning, Grouping, Sudoku
Pictorial Reasoning	Pattern perception, Figure formation, Paper folding and cutting, Figure matrix, Rule detection

Workbook# 10

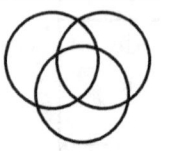

Verbal Reasoning	Arguments-Flawed reasoning, Analogous reasoning
Analytic Reasoning	Positioning, Grouping, Sudoku
Pictorial Reasoning	Pattern perception, Figure formation, Paper folding and cutting, Figure matrix, Rule detection

********* Essential resource for everyone *********
Get the GIFT OF LOGIC™ today !
*http://www.giftoflogic.com   *sales@giftoflogic.com

© Gift Of Logic, Inc * Copying prohibited

Dear Reader:

Your decision to purchase this book is commendable. You now have in your hands, a comprehensive, easy-to-read book in Critical thinking and Logical reasoning that will introduce you to three different areas of thinking and reasoning - Verbal, Analytical and Pictorial. Solving problems in Verbal Reasoning is important to develop a critical mind. Solving problems in Analytic Reasoning is important to develop a flexible and resourceful mind. Solving problems in Pictorial Reasoning is important to develop a visually alert mind.

This book is presented in a workbook format to help you progress quickly. Parents and teachers are urged to complete the exercises ahead of the student and assist them whenever necessary with the help of detailed answers provided at the end of the book. This book can be used as a supplementary resource in the regular class room or it can be used during winter and summer vacations. College/University students, working professionals and retired individuals will also find the Gift Of Logic(tm) Series very useful in enhancing their problem solving abilities, confidence and general intellect.

Critical thinking and Logical reasoning must be practiced consistently to develop strong cognitive skills. After completing the exercises in this book, continue to read the other books in this series to get familiar with different types of Logical reasoning problems.

This workbook is one in a series of twelve workbooks. Please refer to the brochure before this page for a brief description of each workbook. Visit the website http://www.giftoflogic.com for more information.

Happy thinking and reasoning!

TABLE OF CONTENTS

Verbal Reasoning

Assumptions..8
Reasoning Strategy..23

Analytical Reasoning

Sudoku..39
Positioning..44
Grouping...54

Pictorial Reasoning

Patter perception..65
Figure formation...67
Paper folding and cutting...69
Figure matrix..70
Rule detection..72

Answers

Verbal...75
Analytic..103
Pictorial..128

Certificate of Completion

Name _____ Date _____

VERBAL REASONING

Name _____ Date_____

ASSUMPTIONS

In this section on "Assumptions", you will develop the ability to identify the premise that was assumed by an argument to reach its conclusion.

You will be presented with an argument that has a missing premise. The argument leaps to its conclusion without this premise. Your task is to find the missing premise. The missing premise is the assumption that was made to leap to the conclusion. The argument structure is of the following type.

    stated premises + assumed premise => conclusion.

This structure, is shown in its expanded form below:

    stated premise 1
    stated premise 2
    stated premise 3
    missing assumption (premise 4)
    conclusion

After the argument is presented, a question similar to the ones shown below will be asked:
 *Which one of the following assumptions is made by the argument?
 *The argument depends on which one of the following assumptions?

The correct answer is the one that is assumed by the argument to reach its conclusion.

Incorrect answers are those that are irrelevant, and those that do not help the argument reach its conclusion.

Verbal Reasoning
© Gift Of Logic, Inc * Copying prohibited

| 1 | ASSUMPTIONS | Categorical |

There are ten Engineers and ten Politicians in a room. Therefore, it can be concluded that there are twenty people in the room.

Which one of the following premises does the argument assume?

A) No Engineers in the room are Politicians.
B) Some Engineers in the room are Politicians.

2 — ASSUMPTIONS — Conditional

Several companies plant trees throughout the year in a park. All the trees planted and maintained by the GreenScape tree company will grow above 25 feet. Other companies have not been able to achieve this feat. Therefore, the new set of trees that were planted recently in the park will also grow more than 25 feet.

Which one of the following does the argument assume?

A) All the trees that are planted in the park will attain a minimum height of 25 feet.
B) The new set of trees were planted by the GreenScape tree company.
C) The rich nutrients in the park's soil help the trees to grow above 25 feet.

| 3 | ASSUMPTIONS | Maximum |

Students from all the classes in the school will vote to elect their school leader. Whoever gets the maximum number of votes will be the winner. All the students in Roger's class will vote for him. Therefore, he will be elected as the school leader.

Which one of the following does the argument assume?

A) The number of students in Roger's class is more than the number of students in all other classes combined.
B) Roger's class has the most number of students.

4 ASSUMPTIONS — Categorical

Students at the Victory Middle School have the option of taking classes in German or French. This semester, two students enrolled in the French class and two students enrolled in the German class. Therefore, it can be concluded that four different students enrolled in these two classes.

The above argument assumes which one of the following?

A) students who enrolled in German did not enroll in French and students who enrolled in French did not enroll in German.
B) one student enrolled in both French and German classes.

5	ASSUMPTIONS	Symmetry

Children riding bikes that have training wheels find it very useful to learn to balance themselves. The training wheels are fitted into the bike by the bike store which sells them. The two training wheels are fit symmetrically, one on each side. Thus, it is evident that the bike store must be held responsible for any injuries that may be caused by these training wheels.

Which one of the following assumptions does the argument depend on?

A) The training wheels can suddenly become asymmetrical and force the rider to fall off the bike.
B) Stores have special instruments to check the symmetry of the wheels.
C) After several days of use, the nuts that hold the training wheels in place become very tight.

6   ASSUMPTIONS   Causal

Severe thunderstorms accompanied by gusty winds lashed the entire region last night. At the same time, there was power outage in some areas. Therefore, it can be concluded that the thunderstorm caused the power outage in these areas.

The argument assumes which one of the following?

A) The power equipment in the region can withstand even heavier thunderstorms and gusty winds.

B) The areas that had power outage had very old power equipment.

7 ASSUMPTIONS — Conditional

It is not desirable to have a society with quarrelsome people. Therefore, if television programs that encourage quarrels are banned, the society will not have any quarrelsome people.

Which one of the following is an assumption required by the argument?

A) Television programs that encourage quarrels are the main reason for quarrelsome behavior in people.
B) Television programs that show quarrels are the only reason for quarrelsome behavior in people.

8 — ASSUMPTIONS — Medical

Big hospitals have facilities to treat several patients at the same time. The administrators of big hospitals make sure that patients get admitted quickly without much waiting. Therefore, if people have an emergency, they should get admitted into any hospital, big or small.

Which one of the following is an assumption required by the argument?

A) People like to go to big hospitals only in case of an emergency.
B) Small hospitals are as good as big hospitals when it comes to admitting patients quickly.

| 9 | ASSUMPTIONS | Same |

Last month, several passengers in a Cruise ship fell sick suddenly with high fever. Interestingly, the passengers in the same ship have fallen sick suddenly now with high fever. Therefore, the same medicine must be prescribed now as the one that was prescribed last month.

Which one of the following is an assumption required by the argument?

A) The same set of doctors were present in the ship last month and this month when the passengers fell sick.
B) The virus that caused the sickness last month is the same virus that has caused the sickness now.

10 ASSUMPTIONS — Specific

Amar forgot to wear his tie to school. So, his teacher criticized him. Akbar forgot to wear his tie to school. So, Akbar's teacher criticized him. But, when Antony went to school without a tie, his teacher did not criticize him. Therefore, the students are treated fairly by this school.

Which one of the following assumptions does the argument depend on?

A) Kindergarten students like Antony are not required to wear a tie.
B) Antony is in the same class as Amar and Akbar.

11   ASSUMPTIONS   Causal

Computers infected with software categorized as "Virus" cause them to be unusable. Some computer programmers are referred to as "Hackers". Thus, it is clear that "Hackers" are responsible for the unusable computers that we see today.

Which one of the following assumptions does the argument depend on?

A) Even a doctor can be a "Hacker".
B) "Hackers" write "Virus" programs.

12 ASSUMPTIONS — Cost

Sarah's mom gave her ten dollars as pocket money on the first day of each week. Sarah is free to use it as she pleases. So, when her school hosted a book fair, Sarah was able to purchase all the books that were on her wish list with her pocket money.

Which one of the following assumptions does the argument depend on?

A) Sarah saved all her pocket money in order to pay for her purchases at the book fair.
B) The total cost of all the books in her wish list was less than what she could save before the book fair.

13 — ASSUMPTIONS — Compare

Thieves stole the car that John owned and that is why anti-theft features are number one in his priority list when he went shopping for a new car. But, he was in a dilemma regarding which one of two models to purchase. He liked the anti-theft features found in model-A. He also liked the anti-theft features in model-B and decided to purchase a car of this type.

Which one of the following assumptions does the argument depend on?

A) Model-B was cheaper than Model-A.
B) He felt that the anti-theft features of Model-B were difficult to use.

14 ASSUMPTIONS — Highway

People living in two cities, city-A and city-B travel on a congested highway to go to work everyday. If there is a major accident in this highway, the congestion becomes worse. Therefore, in the event of a major accident, it is important to direct the commuters from city-A first to a different route and not the commuters from city-B.

Which one of the following assumptions is made by the argument?

A) It is easier to get on to the highway from city-B than from city-A.
B) A major percentage of commuters using the highway come from city-A.

REASONING STRATEGY

In this section on "Reasoning strategy", you will develop the ability to identify the strategy that is used in an argument. This will help you learn about different argument structures.

An argument is given that has one or more premises and a conclusion. It adopts some strategy to reach its conclusion. After the argument is presented, a question is posed as follows:
* The argument uses which one of the following reasoning methods?
* The argument derives its conclusion by..

The answer choices are presented as follows:
* Jack and Jill use different premises to reach the same conclusion.
* The argument assumes that correlation means causation.
* The argument makes its claim by assuming that the converse of a conditional premise is a valid inference.
* The argument reaches its conclusion by appealing to emotion rather than substance.

The correct answer is the one that identifies the reasoning strategy that is used by the argument. Incorrect answers refer to reasoning strategies that are not present in the argument. The book titled "Primer" discusses argument structure and several types of arguments. The argument may use reasoning strategies that are conditional, causal, categorical, analogous, etc. The argument may be valid or invalid (flawed). Your task is to just identify the strategy that is used by the argument regardless of whether the argument itself is valid or invalid.

1 REASONING STRATEGY — Teenagers

Teenagers are attracted by eye-catching pictures and immediately decide to subscribe to magazines. But, within one year, they stop reading the magazines and cancel their subscriptions. Thus, it is clear that it is not profitable to sell magazines to teenagers.

The argument derives its conclusion by

A) making an assumption that is not stated.
B) presenting all the facts necessary to reach the conclusion.

2 REASONING STRATEGY — Survey

A survey of five hundred men with pot bellies indicates that ninety percent of them drink at least one carbonated drink every day. Carbonated drinks have substances that produce bubbles that cause the stomach to expand. Therefore, carbonated drink is the reason the men surveyed have pot bellies.

The argument employs which one of the following reasoning techniques?

A) It incorrectly applies its conclusion to all the men surveyed, a fact that is true only for ninety percent of them.
B) It uses facts about pot bellied men that were not surveyed to justify its conclusion.

| 3 | REASONING STRATEGY | Travel |

Sonia: Traveling is fun, exciting and adventurous. It is a great way to learn about different types of people and their cultures. Other ways of learning about cultures are boring. Therefore, the best way to learn about culture is by traveling.

Sean: You can learn about world cultures without traveling anywhere. The Encyclopedia software in your computer shows everything about world cultures - food, fashion, etc. Thus, the best way to learn about culture is from the Encyclopedia software.

Which one of the following describes the difference between Sonia's and Sean's arguments?

A) They reach different conclusions using different premises.
B) They reach the same conclusion by citing different premises.
C) They reach their conclusions by using different assumptions.

4 REASONING STRATEGY — Accidents

Recent reports indicate that traffic accidents in the city have increased by twenty five percent. This is very disturbing as it puts everyone at risk, especially the children. These reports also mention that there has been a surge of teenage drivers lately. Therefore, the teenage drivers are responsible for the increase in accidents.

Which one of the following reasoning techniques was used by the argument presented above?

A) It makes an inference by assuming an accurate causal relationship.
B) It makes an inference by assuming an incorrect causal relationship.

5 REASONING STRATEGY — Vending machines

When vending machines were introduced, they dispensed small snack packets like chips and cookies. But, these days, they also dispense lunch items like pizza and sandwich. Therefore, vending machines must stop dispensing snacks from now on.

Which one of the following is true about the reasoning in this argument?

A) It can be logically inferred from the premises.
B) It is not supported by the premises.

6   REASONING STRATEGY   Chemicals

Tom: Mixing the two chemicals together will not cause a major explosion. Therefore, this experiment is safe.

Jerry: I agree that there will not be a major explosion if the two chemicals are mixed together. But still, the experiment is dangerous.

Which one of the following describes how Jerry's argument is structured with respect to Tom's argument?

A) It agrees with Tom's conclusion, but not with the premise.
B) It agrees with Tom's premise, but not with the conclusion.
C) Its conclusion is identical to that of Tom's.

7 REASONING STRATEGY — Transportation

Trains and airplanes offer convenient means of transportation. Trains are safer than airplanes as they operate on the ground. Train tickets are cheaper than airplane tickets due the high cost of maintaining an airplane. But, it will take more time to reach your destination on a train than on an airplane. Thus, since trains have more advantages than airplanes, more money must be spent to improve train facilities instead of airport facilities.

The argument employs which one of the following reasoning techniques?

A) It gives all points of comparison the same weightage.
B) It compares two groups based on items that are biased toward one group.

8 REASONING STRATEGY — Acting

Most of the dramas in which Sanchez acted have been a hit. He loves acting in dramas and has done so since he was seven years old. But, a few dramas that he acted in recently did not do well. So, he is not a popular actor now.

Which one of the following most accurately describes the role played in the argument by the fact that a few dramas that he acted in recently did not do well?

A) It is a counter premise that helps to prove that he does not know how to act.
B) It is a counter premise that helps to prove that he is not a popular actor now.

REASONING STRATEGY — Zoo

Sam: No animal must be kept in a Zoo. Taking them away from their natural habitats and arresting them in a Zoo takes away their right to a natural life. Just because they cannot express their feelings does not mean that they do not have feelings.

Walton: It is quite true that animals have feelings. They must surely be left to live their lives in their original habitats. Nevertheless, without an institution like the Zoo, it is hard for most people to appreciate their existence. Moreover, a Zoo is a great place to seek donations from people for improving their natural habitats. For this reason, keeping a small percentage of animals in a Zoo is more to their advantage than otherwise.

Which one of the following accurately describes Walton's response to Sam's argument that no animal must be kept in a Zoo?

A) It uses a counter premise to present new facts that help reach its conclusion.
B) It agrees with the premise and conclusion in Sam's argument.

10 REASONING STRATEGY — Homework

Principal: Teachers feel that students must not be burdened with lot of homework. They reason that students will spend long hours doing their homework and therefore will not concentrate on their studies during the next day at school. On the other hand, parents feel that students must be given a lot of homework. They reason that without homework, students will waste their time and therefore, keeping them busy with homework is the best way to deal with the problem. Therefore, we must find a solution that works for both parents and teachers.

The reasoning in the argument

A) takes sides with one party's interests while disregarding the other.
B) seeks a solution that would resolve the difference between the two parties.

11 REASONING STRATEGY — Mobile phones

Nowadays, mobile phones come with a camera in them. This camera is very useful for taking pictures of accidents and crimes as they happen. In spite of this, these cameras have been misused by some people who take pictures of others without permission. Therefore, new laws are needed to define the legality of using the cameras in mobile phones.

The fact that cameras in mobile phones have been misused plays which one of the following roles in the argument?

A) It is a premise provided to support the conclusion that cameras found in mobile phones must be banned.
B) It is a premise that supports the conclusion that it is illegal to use mobile phone cameras.
C) It is a counter premise provided to justify the conclusion.

12   REASONING STRATEGY   Magicians

Alexia: Magicians are great. They always do whatever they claim. When they say they can make planes disappear, they actually make it disappear. When they say they can pull a snake out of their empty pockets, they indeed do that.

Alex: Magicians are cheats. They use hypnotism to deceive us. Using hypnotism, they take over our senses and force us to imagine things that they want us to imagine.

Alex counters Alexia's argument by

A) discrediting magicians as cheats because they use hypnotism.
B) claiming that magic and hypnotism are the same.

13 REASONING STRATEGY Toys

Parent: Toy manufacturers must provide computer games as an alternative to plastic and wooden toys. Have you observed the plastic and wooden toys available these days? They come with lots of pieces. While these toys provide some fun initially, they get strewn around the house within minutes, thus making it hard to play with all the pieces at the same time. Moreover, these toys are very small and can be easily swallowed by kids.

The statement that suggests that toy manufacturers must provide computer games, plays which one of the following roles in the argument?

A) It is the conclusion of the argument.
B) It is a sub conclusion of the argument.
C) It is a premise of the argument.

14 REASONING STRATEGY — Mechanic

Mechanic: The chain that connects the two gears in this vehicle is made of copper, a weak material. It is because of this that the chain snapped.

Supervisor: That cannot be true. As you can see, some of the links in the chain have cracks in them. So, it is these cracks that caused the chain to snap.

The supervisor disagrees with the mechanic by

A) making a claim that cannot be inferred from the premises.
B) casting doubt on the mechanic's qualifications.
C) citing a different cause than the one cited by the mechanic.

ANALYTICAL REASONING

1

SUDOKU

Solve the following Sudoku. A correctly solved Sudoku has numbers 1-9 appearing only once in each row, each column and each 3x3 grid. Solving Sudokus will help you to gain valuable analytic skills.

5		4	6	8	7	1	2	
6	9		5		1	4		8
8	7	1		9	2	6	3	5
2	4	7	8		6		5	3
9		5	7	2	3	8		1
1	8	3	9		5	2	6	
7	5	8	2		9	3		4
3	2		1	5	4	7	8	6
4		6	3		8	5	9	2

Analytical Reasoning Answers-103

2
SUDOKU

Solve the following Sudoku. A correctly solved Sudoku has numbers 1-9 appearing only once in each row, each column and each 3x3 grid. Solving Sudokus will help you to gain valuable analytic skills.

	4	7		3	5	1	8	2
1	6	3	2		9	5	4	7
8		2	7	1	4		3	9
5	2	9	8		1	4	7	3
3		8		4		2	6	5
6	7	4	3	5	2		9	1
7	8	6	5	2	3	9		4
2		1	4		6	7		8
4	9	5	1	7	8	3	2	6

Analytical Reasoning   Answers-104

3

SUDOKU

Solve the following Sudoku. A correctly solved Sudoku has numbers 1-9 appearing only once in each row, each column and each 3x3 grid. Solving Sudokus will help you to gain valuable analytic skills.

7		9	2	1	6	5	8	4
2	1		7		8		3	
5	8	6	4	9	3	2	7	1
6	5	1	3			2	4	9
9		8	1		5	6	2	3
4	2	3	6	7	9	8	1	5
1	6	2	9	3	4	7	5	8
8		7	5			1	3		2
3	4	5	8	2	7	1	6	9

Analytical Reasoning   Answers-105

4
SUDOKU

Solve the following Sudoku. A correctly solved Sudoku has numbers 1-9 appearing only once in each row, each column and each 3x3 grid. Solving Sudokus will help you to gain valuable analytic skills.

3	8	9	1	7	6	4	2	5
2		7	5	8	9			6
5	1	6	4	3	2	7	8	9
1	2		7	4	5	9		8
9	5	4	8	6			7	3
7	6			2	3	5	4	1
8	7	5	3		4	6		2
4	3	2		9	8	1	5	7
6		1	2		7		3	4

Analytical Reasoning Answers-106

Name —————————————— Date ——————————

5
SUDOKU

Solve the following Sudoku. A correctly solved Sudoku has numbers 1-9 appearing only once in each row, each column and each 3x3 grid. Solving Sudokus will help you to gain valuable analytic skills.

9		2	7	8	3	6	4	1
8	3	6	1		5	7	9	2
	7	1	9	2	6	3	8	5
5	2	3	6	9	4		7	8
7	8	4		1		9	5	6
1	6	9	8	5	7	2		4
2	1	8	4	3	9	5	6	7
3	4	7	5		1	8		9
6	9		2	7	8	4	1	3

Analytical Reasoning

1   POSITIONING   can be true, if-then

SCENARIO

Andrea, Bobby, and Cindy will sing consecutively in a competition. If Andrea sings first, then Bobby cannot sing next.

QUESTIONS

Which of the following can be true?

A) Andrea sings first and Cindy sings last.
B) Cindy sings first and Bobby sings next.
C) Andrea, Cindy, and Bobby sing one after the other.
D) Bobby, Andrea, and Cindy sing consecutively.

2 POSITIONING — cannot be true

SCENARIO

Chris, Raj, and Walid will perform consecutively in a piano recital. Walid cannot be the last one to perform.

QUESTIONS

Which of the following cannot be true?

A) Walid, Raj, and Chris play first, second and third respectively.
B) Raj plays first, followed by Chris.
C) Chris plays first, followed by Raj.
D) Walid, Chris, and Raj play one after the other in that order.

3 POSITIONING   vertical position, can be true

SCENARIO

Three clowns, Mumbo, Jumbo, and Fumbo must stand consecutively in the steps of a three-step ladder.

The steps are numbered from 1 to 3 from the bottom.

Jumbo must always be above Mumbo.

QUESTIONS

1) Which of the following can be true?
    A)   Fumbo is in position 2.
    B)   Jumbo is in position 1.

2) Which of the following cannot be true?
    A)   Mumbo is in 1st position.
    B)   Mumbo is in 2nd position.
    C)   Mumbo is in 3rd position.

4 POSITIONING — logical chaining

SCENARIO

Andrea, Brianna, and Cindy go to watch a movie. They want to sit together in three seats, consecutively.

They must sit in the order of their heights.
Andrea is taller than Brianna.
Brianna is taller than Cindy.

QUESTIONS

1) List all the possible ways that Andrea, Brianna, and Cindy can sit.

2) Of the possible ways that the three can be seated, which of the following is always true?

    A) Brianna sits between Andrea and Cindy.
    B) Cindy sits in the right most seat.
    C) Andrea sits in the left most seat.

5   POSITIONING   logical chaining

SCENARIO

Vivek, Brendon, and Farhad will perform in a piano music festival along with several others. The performances will be scheduled one after the other.

QUESTIONS

1) If Vivek must play before Farhad and Brendan must play after Farhad, then which of the following orders is possible?

    A) Vivek, Farhad, Brendan, John
    B) Vivek, John, Brendan, James, Farhad
    C) Farhad, Brendan, John, Vivek

2) In the festival, if Vivek plays before Farhad and Brendan plays after Vivek, then which of the following orders is possible?

    A) Brendan, Vivek, Farhad
    B) Vivek, Brendan, Farhad
    C) Vivek, Farhad, Brendan

Analytical Reasoning   Answers-112

Name _____   Date _____

6   POSITIONING   fixed position

SCENARIO

Rudy, Puppy, Tommy, and Tony are four dogs that need to be seated in four consecutive spots.

Rudy must always sit in the third spot.
Tommy and Tony must always sit in consecutive positions.

QUESTIONS

1) In which of the following positions can Puppy be seated?

    A) first   B) second   C) third   D) fourth

2) Which of the following is always true?

    A) Puppy always sits in the fourth position.
    B) Tony always sits in the first position.
    C) Tommy always sits in the second position.

3) If Rudy has to sit in the second spot instead of the third spot, then Puppy can sit in which of the following spots?

    A) first B) second C) third D) fourth

Analytical Reasoning   Answers-113

7 POSITIONING — either or

SCENARIO

Three mice get ready to sneak into a kitchen to eat cheese. They agree to stay together one behind the other.

Their names are Rambo1, Rambo2, and Rambo3.
Either Rambo1 or Rambo2 must be in the middle.

QUESTIONS

1) If Rambo1 is in the second position, what is the position of Rambo2?

2) If Rambo3 is in the first position, then Rambo1 must be in the second position.

    A) True   B) False

3) Is the following order of the mice possible as they sneak into the kitchen?

    Rambo1, Rambo3, Rambo2

    A) Yes   B) No

8 POSITIONING — at least, lowest, highest

SCENARIO

Three boxes labeled as Golf balls, Tennis balls, and Ping-pong balls respectively, are to be stacked one over the other. The bottom position is position #1. There should be at least one box below the boxes labeled "Tennis balls" and "Ping-pong balls"

QUESTIONS

1) What is the lowest position in which the box labeled 'Tennis balls' can be placed?   A) 1   B) 2   C) 3

2) What is the highest position in which the box labeled 'Tennis balls' can be placed?   A) 1   B) 2   C) 3

3) What is the lowest position in which the box labeled 'Ping-pong balls' can be placed?   A) 1   B) 2   C) 3

4) What is the highest position in which the box labeled 'Golf balls' can be placed?   A) 1   B) 2   C) 3

Name _____   Date _____

9 POSITIONING every other

SCENARIO

George plays soccer Monday through Friday for four weeks every month. During the first and third weeks, he is the captain of the team every other day starting from Monday. During the second and fourth weeks, he is the captain every other day starting from Tuesday. Use the table below to answer the questions.

Week	Monday	Tuesday	Wednesday	Thursday	Friday
1					
2					
3					
4					

QUESTIONS

1) For at most how many days in the month can George be the captain of the team?
    A) 6   B) 10

2) For at most how many days in the second week is someone other than George the captain of the soccer team?
    A) 2   B) 3

3) George captains more number of days during the first week than he does on the third week.
    A) True   B) False

Name _____   Date _____

10 POSITIONING   ranking position

SCENARIO

The scores of three students in the Science, Math, and English exams are as follows:

Subjects	Student 1	Student 2	Student 3
Science	90	95	80
Math	80	70	90
English	70	60	60
Total -All			
Total -Math and Science			

QUESTIONS

Fill in the blank cells in the table and answer the questions below.

1) Based on the total scores, the ranking of students is
    A)   Student 1, Student 2, Student 3
    B)   Student 1, Student 3, Student 2
    C)   Student 3, Student 2, Student 1

2) The ranking of students based on their math and science scores is
    A)   Student 1, Student 3, Student 2
    B)   Student 1, Student 2 and Student 3
    C)   Student 1 and Student 3, Student 2

3) If Student 2 had scored 76 in Math, then the rank of Student 1 based on Math and Science totals would be which one of the following?
    A) 1   B) 2   C) 3

Analytical Reasoning   Answers-117

1 GROUPING — at least one, if-then

SCENARIO

A sales manager sells his products in three cities P, Q, and R. In his trips, he must visit at least one city. If he does not visit city P, then he must not visit city Q.

QUESTIONS

1) Which of the following cities can the sales manager visit?
   A) Q and R
   B) P and R
   C) P and Q
   D) P, Q, and R
   E) P
   F) Q

Analytical Reasoning   Answers-118
© Gift Of Logic, Inc * Copying prohibited

2 GROUPING — at least three, if-then

SCENARIO

Danny went to a dentist to have his teeth cleaned. The dentist identified four decayed teeth T1, T2, T3, and T4. He needs to pull out at least three of them. If tooth T1 is not pulled out, tooth T3 will not be pulled out.

QUESTIONS

1) Which of the following sets of teeth can be pulled out of Danny's mouth?

    A) T1, T2, T3
    B) T2, T3, T4
    C) T1, T2, T4

3 GROUPING — several conditions

SCENARIO

Seeds of five different vegetables are available to be planted in a garden. At least three of them must be planted.

If veggie-1 and veggie-2 are planted, then veggie-4 must not be planted. If veggie-2 and veggie-3 are planted, then veggie-5 must be planted. If veggie-5 is not planted, then veggie-4 must not be planted as well.

QUESTIONS

1) Which of the following groups of veggies can be planted in the garden.

    A) veggie-2, veggie-3, veggie-5
    B) veggie-2, veggie-3, veggie-4
    C) veggie-2, veggie-3, veggie-1, veggie-4
    D) veggie-1, veggie-2, veggie-3, veggie-4, veggie-5

Name _____   Date _____

4   GROUPING   some, but not all

SCENARIO

Andrew has a choice of eating three different ice-creams, IC-1, IC-2, and IC-3. He can eat some, but not all the ice-creams. He cannot eat IC-3 and IC-1 together.

QUESTIONS

1) Which of the following represent the different combinations of ice creams that Andrew can eat?
     A) IC-1, IC-3
     B) IC-2, IC-3
     C) IC-3, IC-1, IC-2
     D) IC-3, IC-1
     E) IC-2

5   GROUPING   at most, at least one of

SCENARIO

There are five theme parks, P1, P2, P3, P4, and P5 in Marvelous city. Paula has the time only to visit at most three of the parks. At least one of P2 and P3 must be visited.

QUESTIONS

1) Which of the following represent the different theme parks that Paula can visit?
     A) P1, P4, P5
     B) P2, P4, P5
     C) P4, P3, P2

| 6 | GROUPING | at least, but not all |

SCENARIO

Movies m1, m2, and m3 are shown in the mornings and movies m4 and m5 are shown in the afternoons. Neil must watch at least three, but not all the movies. He must see at least two movies in the morning and at least one in the afternoon.

QUESTIONS

1) Which of the following choices does Neil have to watch movies?

    A) m1, m2, m3, m4
    B) m1, m4, m5
    C) m1, m3, m4, m5
    D) m1, m2, m3
    E) m1, m3

Analytical Reasoning   Answers-122

Name _____   Date _____

7 GROUPING   minimum, maximum

SCENARIO
A minimum of five students and a maximum of ten students must be selected from a pool of twenty five students.

QUESTIONS
1) What is the maximum number of students that will miss the selection?

2) What is the minimum number of students what will miss the selection?

8 GROUPING   minimum, must, must not

SCENARIO
A team comprising of a minimum of four students must be picked from a group of six students A, B, C, D, E, and F. A and B must be picked together. E and F must not be picked together. B and E must be picked together.

QUESTIONS
1) Which of the following set of students can be picked?
   A) A, B, E, F
   B) A, B, C, E
   C) A, B, C, D
   D) B, E, C, D
   E) C, D, E
   F) B, C, F

9 GROUPING — only, multiple groups

SCENARIO

Four people are selected from two groups of four people each. There should be only two selected from each group. Group 1 has A, B, C, D and Group 2 has E, F, G, H. If A is selected, then G and H must be selected together. If D is selected, then E and F must be selected together.

QUESTIONS

1) If A is selected from group-1, then which of the following can also be selected?
    A) B   B) D

2) If D is selected from group-1 and B is not selected, then which of the following can also be selected?
    A) A   B) C

Name _____   Date _____

10

GROUPING   multiple groups, at least

SCENARIO

A, B, C, and D are members of Group P.
E, F, G, and H are members of Group Q.
I, J, K, and L are members of Group R.

Four members must be selected from these three groups. At least one member must represent each of the three groups.

QUESTIONS

1) If B, G, and K are selected and there should not be two members selected from Group P, then the fourth member can be from which of the following sets?

    A) (E,F,H) or (I,J,L)
    B) (A,C,D)
    C) (A,C,K)

2) If B, I, and K are selected and there should not be two from either P or Q, then the fourth member can be from which of the following sets?

    A) Q
    B) (A,C,D)
    C) (J, L)

Analytical Reasoning   Answers-125
© Gift Of Logic, Inc * Copying prohibited

Name —————————————— Date——————

11
GROUPING   multiple groups, group size

SCENARIO

L, M, N, and O are members of Group P.
R, S, T, U, and V are members of Group Q.

A team of four must be selected from groups P and Q.
There should be more members from group P than from group Q.
At least one from P and at least one from Q must be in the team.

QUESTIONS

1) How many members can be selected from P and Q?
   A) three from P, one from Q
   B) two from each
   C) one from P, three from Q
   D) four from Q

2) If L and R are selected into the team, then the remaining two will come from which of the following groups?

   A) (LMNO) and (RSTUV)
   B) (MNL) and (RTUV)
   C) (ONM)

Analytical Reasoning   Answers-126
© Gift Of Logic, Inc * Copying prohibited

Name _____   Date _____

12   GROUPING   multiple groups

SCENARIO

L, M, N, and O are members of Group P.
R, S, T, U, and V are members of Group Q.
A team comprising of members from both Group P and group Q is to be formed.

1) If there should be as many members in the team from group P as there are from group Q, then which of the following represents a valid team?

     A) L,M,N,R,S   B) L,M,N,R,S,U
     C) T,U,V,L,M   D) L,T

13   GROUPING   sub group, group size

SCENARIO

L, M, N, O are members of Group P.
R, S, T, U are members of Group Q.

Sub groups Y and Z, comprising of three members each must be formed. At least one member of each subgroup must be from P and one member must be from Q. In sub-group Y, there should be more members from P than from Q. In sub-group Z, there should be more members from Q than from P.

1) How many members are there in a group that is formed by combining the members of groups Y and Z?

     A) three from P, three from Q
     B) three from P, four from Q

Analytical Reasoning   Answers-127

Name —————————— Date ——————————

PICTORIAL REASONING

PATTERN PERCEPTION - MISSING PATTERN

Find the correct figure from the two alternatives given that will fit logically into the missing portion of the figure on the left.

1

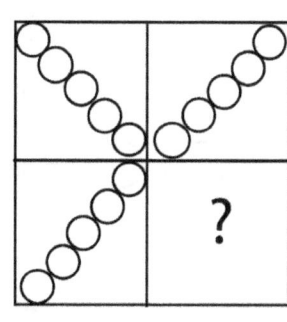   A 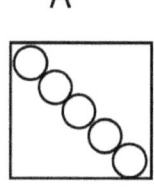   B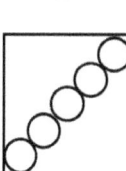

2

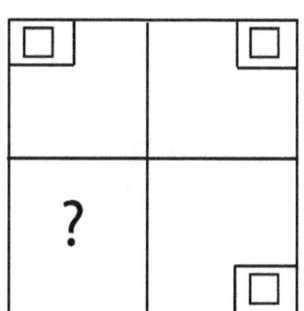   A 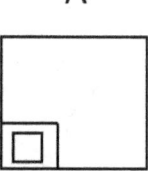   B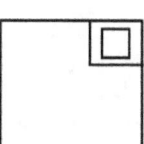

3

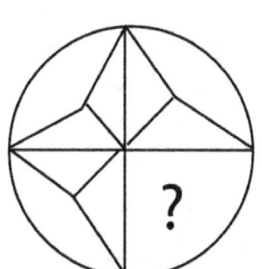   A 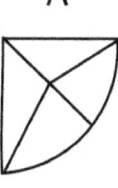   B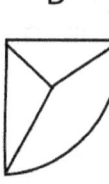

4

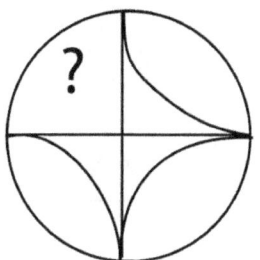   A    B

Pictorial Reasoning   Answers-128   65
© Gift Of Logic, Inc * Copying prohibited

Name ——————————————————   Date——————————————

PATTERN PERCEPTION - CONTINUING PATTERN

Find the correct figure from the two alternatives given that will logically continue the pattern of figures on the left.

5

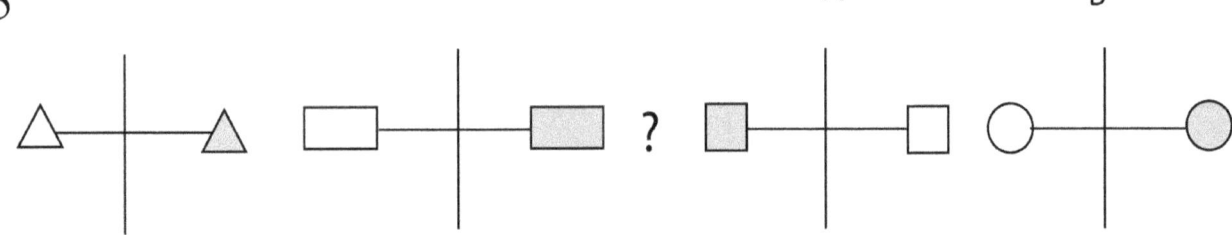

6

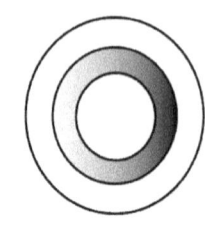 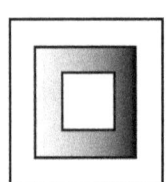 ?

7

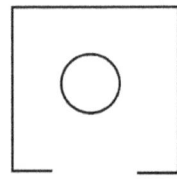 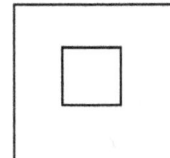 ?

8

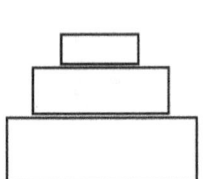  ? A B

Name _____   Date _____

FIGURE FORMATION

Find the correct figure that will be formed when the figures on the left are combined. Either of the figures may be rotated before combining.

1   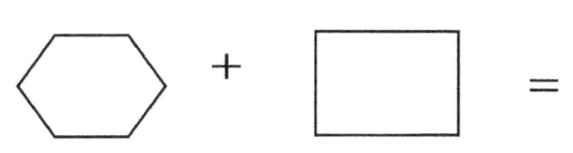   A 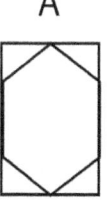   B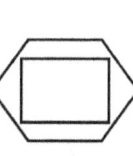

2   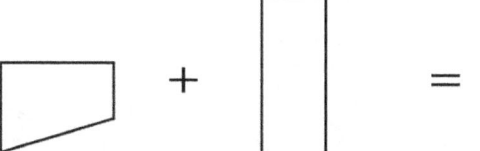   A 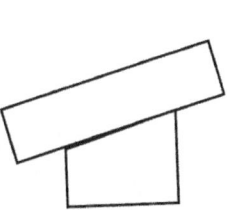   B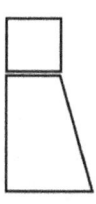

3   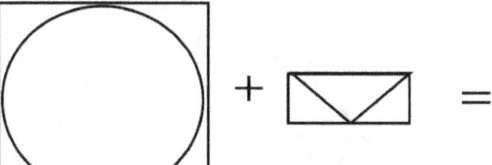   A 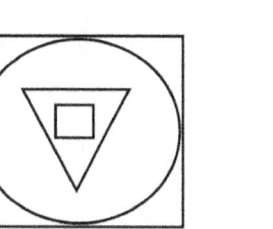   B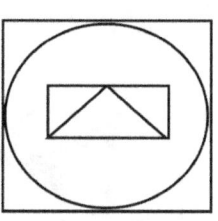

4   A 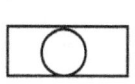   B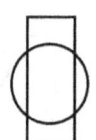

Pictorial Reasoning   Answers-128
© Gift Of Logic, Inc * Copying prohibited

FIGURE FORMATION

Find the correct figure that will be formed when the two figures on the left are combined. Either of the figures may be rotated or scaled before combining.

5

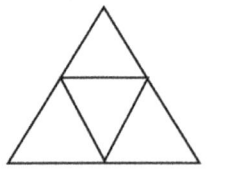 + 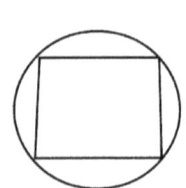 =   A   B

6

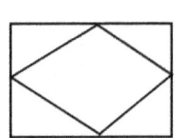 + 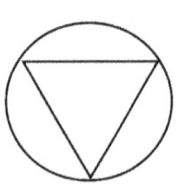 =   A   B
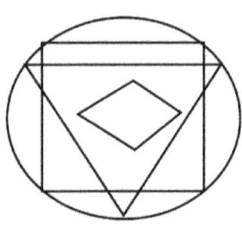   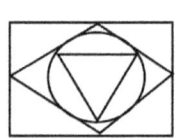

7

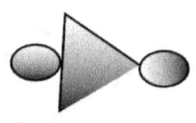 + 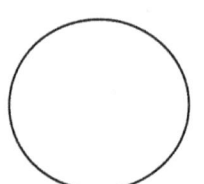 =   A   B

8

 + ○ =   A   B

Pictorial Reasoning   Answers-128
© Gift Of Logic, Inc * Copying prohibited

PAPER FOLDING AND CUTTING

Find the correct figure that will be formed when the paper on the left is folded in the direction of the arrows, and then holes are cut in it as shown.

1	A	B	C	D
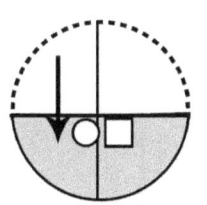	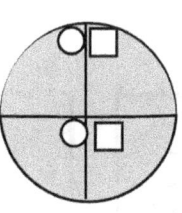	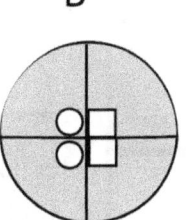	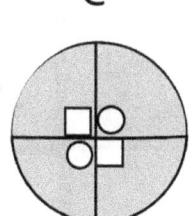	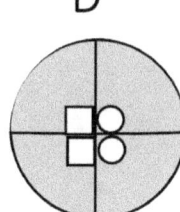

2	A	B	C	D
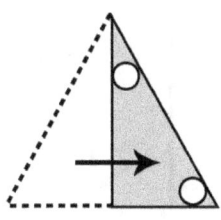	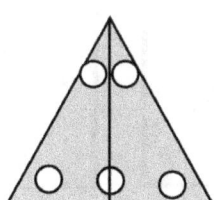	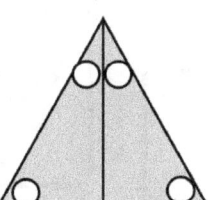	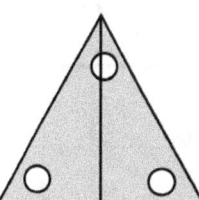	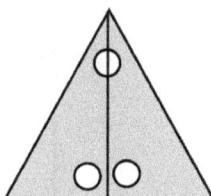

3	A	B	C	D
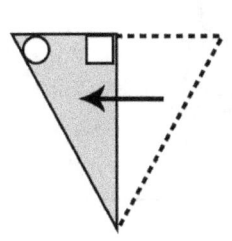	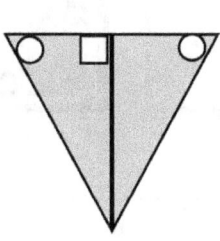	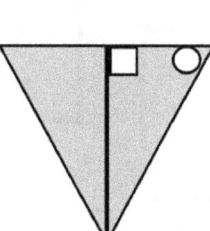	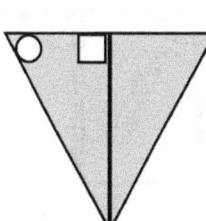	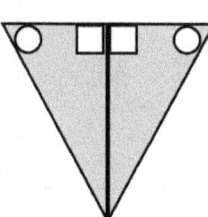

Pictorial Reasoning
© Gift Of Logic, Inc * Copying prohibited

FIGURE MATRIX - ANALOGY

Find the correct figure from the alternatives given that will fit in the empty box such that, the bottom two figures are related in the same way as the top two figures.

1

2

3

4

Pictorial Reasoning   Answers-129

Name _____ Date _____

FIGURE MATRIX- SIMILARITY

Three figures in the 2 x 2 matrix have similar characteristics. Find the fourth figure from the alternatives given that is also alike.

5 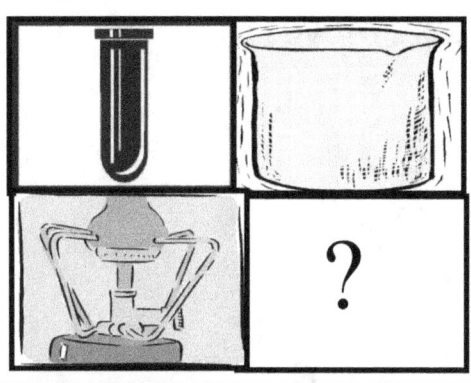   A    B    C

6 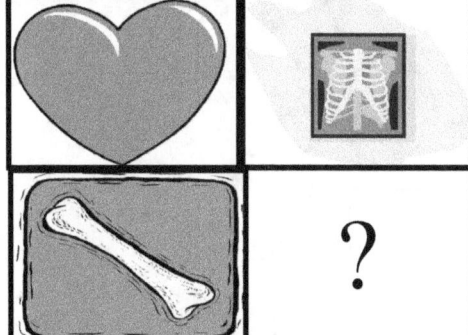   A 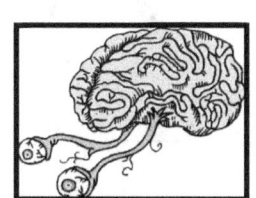   B 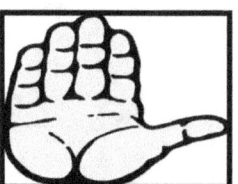   C

7    A 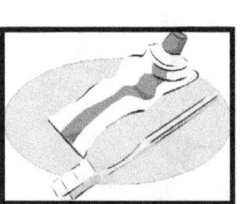   B    C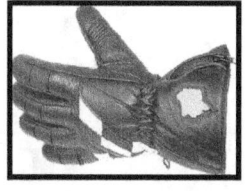

8    A    B    C

Pictorial Reasoning   Answers-129
© Gift Of Logic, Inc * Copying prohibited

Name —————————————— Date ——————————

RULE DETECTION

Read the given rule in each question. Then, find the correct choice from the alternatives given that satisfies the rule.

The circle moves from right to left while the square moves from left to right

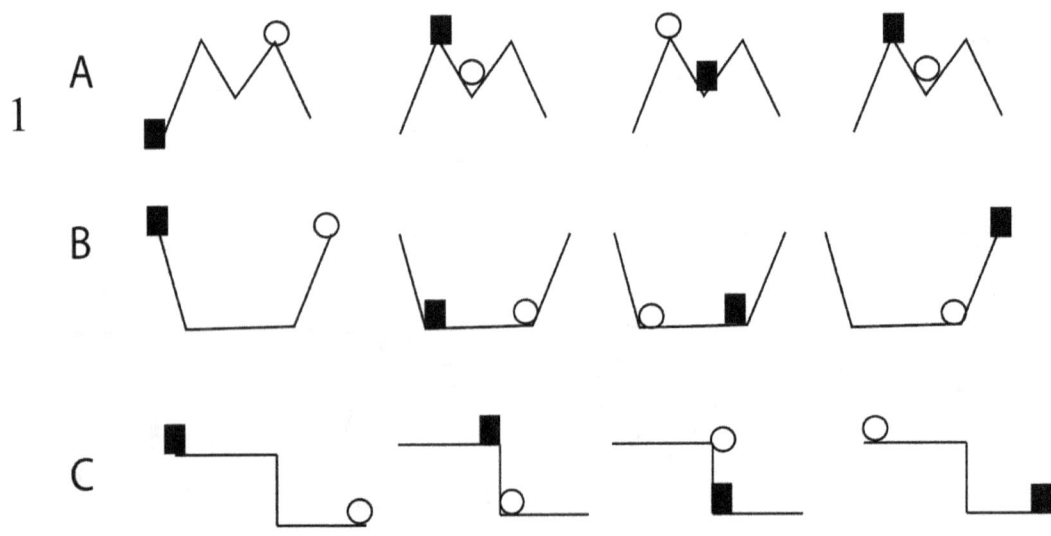

1 A

   B

   C

———

The circles move clockwise; the squares move counter clockwise

2 A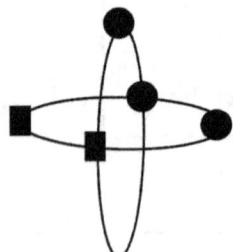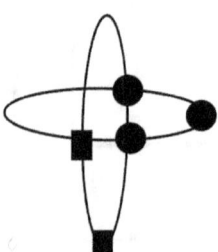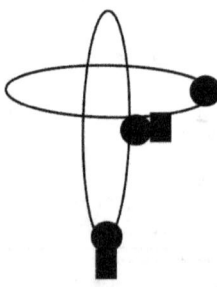

   B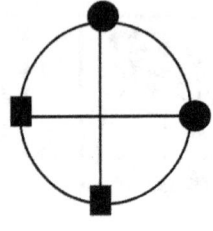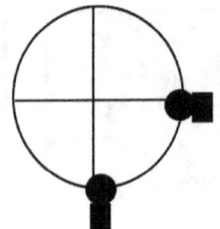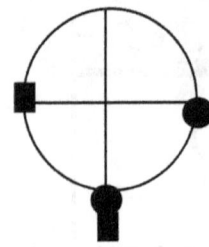

Pictorial Reasoning
© Gift Of Logic, Inc * Copying prohibited

Name _____   Date _____

RULE DETECTION

Read the given rule in each question. Then, find the correct choice from the alternatives given that satisfies the rule.

Figures in bottom row are reflections of figures in top row

3

A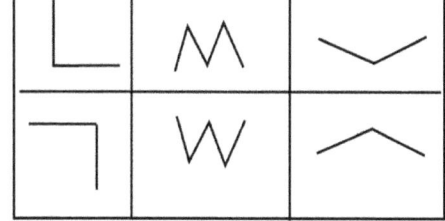

B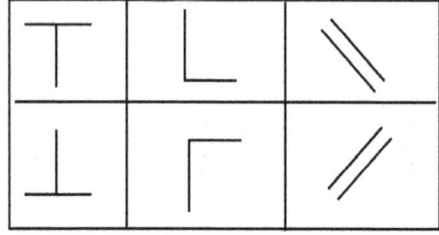

Letters in right column are reflections of letters in left column

4   A   B

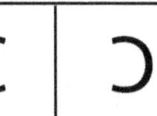   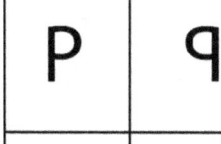

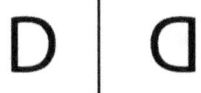

Pictorial Reasoning   Answers-129   73
© Gift Of Logic, Inc * Copying prohibited

ANSWERS

1 ASSUMPTIONS

There are ten Engineers and ten Politicians..
Which one of the following does the argument assume?
A) No Engineers in the room are Politicians.
B) Some Engineers in the room are Politicians.

ANSWER

Answer: A

A – correct - if no Engineers are Politicians, this also means that no Politicians are Engineers – so, ten Engineers plus ten Politicians will equal twenty people in the room, which is the conclusion. See Venn diagram below.

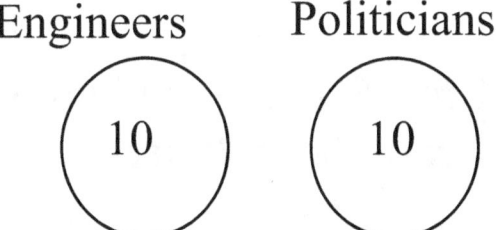

B – incorrect- this assumption states that some Engineers are also Politicians. Assume that one person is both and Engineer and a Politician. This means that nine are Politicians, but not Engineers and nine are Engineers, but not Politicians and one is both. Adding all the people, we get nineteen people only, less than the twenty that must be in the room. The Venn diagram below shows this scenario. So, when some Engineers are also politicians, we cannot have twenty people in the room.

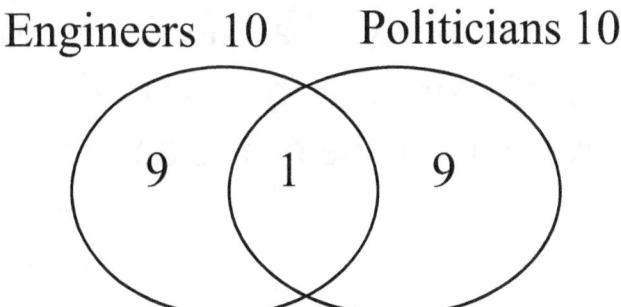

Answers
© Gift Of Logic, Inc * Copying prohibited

2 ASSUMPTIONS

Several companies plant trees..

Which one of the following does the argument assume?
A) All the trees that are planted in the park will attain a minimum height of 25 feet.
B) The new set of trees were planted by the GreenScape tree company.
C) The rich nutrients in the park's soil help the trees to grow above 25 feet.

ANSWER

Answer: B

premise: if the tree is planted by GreenScape → will grow above 25 feet
conclusion: new set of trees will grow above 25 feet

A – incorrect - this answer choice contradicts the premise that only the trees planted by GreenScape grow above 25 feet. The assumption contradicts an existing premise and does not lead to the stated conclusion.

B – correct – since the conclusion is that the new set of trees will grow above 25 feet, this can be possible only if the condition is satisfied, that is, the new set of trees were planted by GreenScape.

C – incorrect – this assumption says that the rich nutrients help the trees (all trees) to grow above 25 feet. But, this contradicts the stated fact that only the trees planted by GreenScape grows above 25 feet.

Answers
© Gift Of Logic, Inc * Copying prohibited

3	ASSUMPTIONS

Students from all the classes will elect..
Which one of the following does the argument assume?
A) The number of students in Roger's class is more than the number of students in all other classes combined.
B) Roger's class has the most number of students.

ANSWER

Answer: A
  premise: Students from all the classes will vote to elect a school leader.
  premise: Whoever gets the most votes will win.
  premise: All the students in Roger's class will vote for him.
  conclusion: Therefore, Roger will be elected as the school leader.

A – correct - Since, everyone in Roger's class will vote for him, even if everyone in other classes do not vote for him, he will have the most votes only if there are more students in his class than all other classes combined. For example, if there are 100 students in his class (all 100 will vote for him) and 99 students in all other classes combined, even if all the 99 students vote against him, he will still win the election because he has one extra vote to win.

B – incorrect – even if his class has the most number of students compared to other classes, it is still possible that there are more number of students in all other classes combined. For example, if Roger's class has 100 students and two other classes had 60 and 50 students respectively, that would mean that there are 110 students who are not in Roger's class. If they all vote against him, he will not be elected. So, this assumption is incorrect.

Answers

4 ASSUMPTIONS

Students at the Victory Middle School have..

The above argument assumes which one of the following?
A) students who enrolled in German did not enroll in French and students who enrolled in French did not enroll in German.
B) one student enrolled in both French and German classes.

ANSWER

Answer: A

premise: two students enrolled in French
premise: two students enrolled in German
conclusion: four different students enrolled in these two classes
A – correct - this assumption ensures that no student was counted twice and validates the conclusion that there were four different students enrolled in these two classes.

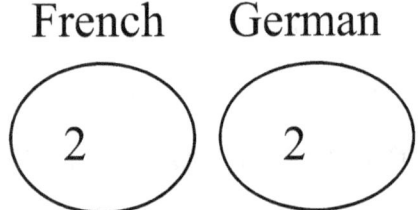

B – incorrect - if this assumption was true, then there would be a total of only three different students enrolled in French and German classes. See Venn diagram below. This assumption does not lead us to the conclusion.

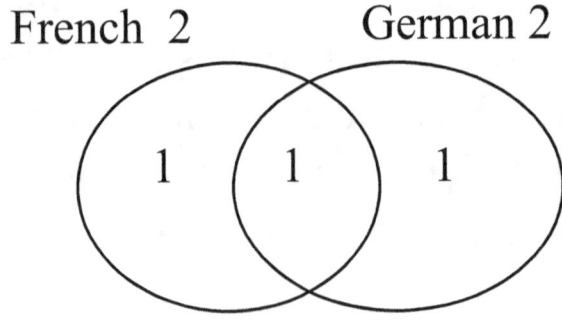

Answers
© Gift Of Logic, Inc * Copying prohibited

5 ASSUMPTIONS

Children riding bikes that have training..

Which one of the following assumptions does the argument depend on?
A) The training wheels can suddenly become asymmetrical and force the rider to fall off the bike.
B) Stores have special instruments to check the symmetry of the wheels.
C) After several days of use, the nuts that hold the training wheels in place become very tight.

ANSWER

Answer: A

premise: bike store fits the training wheels symmetrically.
conclusion: bike stores are responsible for injuries caused by these wheels.

A – correct – if the training wheels can suddenly become asymmetrical and force the rider off the bike, this premise would help justify the conclusion that bike stores are to be held responsible for injuries caused by training wheels.

B – incorrect – this statement does not support the argument that they must be held responsible for injuries caused by the bike.

C – incorrect – if this is the case, then the training wheels will not cause the rider to fall off and sustain injuries. So, this assumption is not correct.

Answers

6 ASSUMPTIONS

Severe thunderstorms accompanied..

The argument assumes which one of the following?
A) The power equipment in the region can withstand even heavier thunderstorms and gusty winds.
B) The areas that had power outage had very old power equipment.

ANSWER

Answer: B

premise: severe thunderstorms & gusty winds lashed the entire region
premise: electricity outage in some areas at the same time
conclusion: thunderstorm caused the power outage

A - incorrect – if this assumption was correct, then there would not have been any power outage. But, clearly, there was a power outage in some areas. So this assumption is incorrect.

B - correct – if this assumption was true, then the old power equipment could not have withstood the thunderstorm. This would explain the conclusion that the thunderstorm caused the power outage in certain areas.

7 ASSUMPTIONS

It is not desirable to have a society with..

Which one of the following is an assumption required by the argument?
A) Television programs that encourage quarrels are the main reason for quarrelsome behavior in people.
B) Television programs that show quarrels are the only reason for quarrelsome behavior in people.

ANSWER

Answer: B

premise: not desirable to have a society with quarrelsome people
conclusion:
   ban TV programs that encourage quarrels → no quarrelsome people

A – incorrect – if TV programs that encourage quarrels are the main reason (but not the only reason) for society having quarrelsome people, then there may be other reasons that cause quarrelsome behavior as well. So, banning TV programs alone will not remove quarrelsome people. This assumption is therefore incorrect.

B – correct – if TV programs showing quarrels are the only reason for quarrelsome behavior, banning the TV programs will result in a society without quarrelsome people. Therefore, this assumption is correct.

8 ASSUMPTIONS

Big hospitals have facilities to treat several patients..

Which one of the following is an assumption required by the argument?
A) People like to go to big hospitals only in case of an emergency.
B) Small hospitals are as good as big hospitals when it comes to admitting patients quickly.

ANSWER

Answer: B

premise: big hospitals can treat several patients at the same time
premise: can get admitted quickly into big hospitals without much waiting
conclusion: if people have an emergency, they get admitted to a big or small hospital

The premises in the argument are about big hospitals, whereas the conclusion leaps to recommend small hospitals as well.

A – incorrect – if this were the case, then the conclusion that people should get admitted to small hospitals as well would be wrong.

B – correct – this assumption, that small hospitals are as good as big hospitals, provides the necessary evidence to support the conclusion that people should get admitted to small hospitals or big hospitals in case of an emergency.

Answers
© Gift Of Logic, Inc * Copying prohibited

9 ASSUMPTIONS

Last month, several passengers in a Cruise ship..

Which one of the following is an assumption required by the argument?
A) The same set of doctors were present in the ship last month and this month when the passengers fell sick.
B) The virus that caused the sickness last month is the same virus that has caused the sickness now.

ANSWER

Answer: B

premise: last month, several passengers fell sick suddenly with high fever
premise: passengers in the same ship are sick now with high fever
conclusion: same medicine that was prescribed last month must be prescribed now

A – incorrect – just because the same set of doctors were present in the cruise ship last month and this month does not mean that the same medicine must be prescribed in both cases. The cause of the sickness may be different in both cases. So, this statement does not support the argument.

B – correct – since the same virus caused the sickness both times, the doctors must prescribe the same medicine now as they did last month.

Answers
© Gift Of Logic, Inc * Copying prohibited

10 ASSUMPTIONS

Amar wore his school uniform, but forgot..

Which one of the following assumptions does the argument depend on?
A) Kindergarten students like Antony are not required to wear a tie.
B) Antony is in the same class as Amar and Akbar.

ANSWER

Answer: A

premise: Amar did not wear a tie and got criticized.
premise: Akbar did not wear a tie and got criticized.
premise: Antony did not wear a tie, but did not get criticized.
conclusion: Students are treated fairly by this school.

A – correct - this explains why Antony was not criticized. A specific information about Antony (that he is a kindergarten student and that kindergarten students are not required to wear a tie) was assumed by the argument to reach its conclusion. Thus, all the three students were treated fairly. This statement supports the argument.

B – incorrect – if this is the case, why was Antony not criticized? So, this assumption does not support the conclusion of the argument.

11. ASSUMPTIONS

Computers infected with software categorized..

Which one of the following assumptions does the argument depend on?
A) Even a doctor can be a "Hacker".
B) "Hackers" write "Virus" programs.

ANSWER

Answer: B.

    causal premise: Virus c→ unusable computers
        premise: Hackers are computer programmers
    conclusion: Hackers are responsible for the unusable computers

From the above, it is clear that an assumption is made linking the Hackers and Virus programs.

A – incorrect – this assumption does not link the Hacker to the Virus program.

B – correct - this assumption links the Hackers and the Virus programs. It indicates that Hackers write the Virus programs. Summing up all the premises and conclusion, we get the following:

    Virus programs c→ unusable computers
    Hackers write the Virus programs.
    Therefore, Hackers are responsible for the unusable computers.

The symbol c→ is used to represent causal relationships. Refer to the Primer for detailed discussion on causal relationships.

Answers

12 ASSUMPTIONS

Sarah's mom gave her ten dollars ..

Which one of the following assumptions does the argument depend on?
A) Sarah saved all her pocket money in order to pay for her purchases at the book fair.
B) The total cost of all the books in her wish list was less than what she could save before the book fair.

ANSWER

Answer: B

premise: Sarah's mom gave her ten dollars each week.
conclusion: Sarah bought all the books in her wish list.

A - incorrect - even if she saved all her pocket money, this does not explain if that would be sufficient to pay for her wish list at the book fair.

B - correct - the total cost of all the books in her wish was less than the savings. So, she was able to purchase all the books in her wish list with the money that she saved.

Answers
© Gift Of Logic, Inc * Copying prohibited

13 ASSUMPTIONS

Thieves stole the car that John owned..

Which one of the following does the argument depend on?
A) Model-B was cheaper than Model-A.
B) He felt that the anti-theft features of Model-B were difficult to use.

ANSWER

Answer: A

premise: thieves stole John's car
sub conclusion: so, he wanted anti-theft features in new car
premise: confused whether to buy model-A or model-B
premise: liked the anti-theft features in model-A
premise: liked the anti-theft features in model-B
conclusion: decided to purchase model-B

A – correct – John liked the anti-theft features in both models. So, he could have bought either model. But, he decided to buy a model-B car. So, the assumption that "model-B was cheaper than model-A" fills the gap in premises and helps support his decision to buy a car of model-B.

B – incorrect – if this were the case, he would not have purchased Model-B, but he did. So, this is not a valid assumption that would lead us to the conclusion.

14 ASSUMPTIONS

People living in two cities, city-A and city-B ..

Which one of the following assumptions is made by the argument?
A) It is easier to get on to the highway from city-B than from city-A.
B) A major percentage of commuters using the highway come from city-A.

ANSWER

Answer: B

premise: people of city-A and city-B use the same congested highway
premise: accidents make the congestion worse
conclusion: in case of major accident, commuters from city-A must be directed to a different route first.

A – incorrect – if it easier to get on to the highway from city-B, then in case of a major accident, the commuters from city-B must be directed to a different route. But, that is not the conclusion. So, this assumption is incorrect.

B – correct – if a major percentage of commuters come from city-A , as this choice suggests, then it would make sense to direct commuters from city-A to a different route. This assumption supports the conclusion and is the correct answer.

1 REASONING STRATEGY

Teenagers are attracted by eye-catching pictures..

The argument derives its conclusion by
A) making an assumption that is not stated.
B) presenting all the facts necessary to reach the conclusion.

ANSWER

Answer: A

> teenagers cancel their subscription within one year.
> thus, it is not profitable to sell magazines to teenagers.

A- correct

Since teenagers cancel their subscription within one year, the argument concludes that it is not profitable to sell magazines to them. The argument leaps to its conclusion by assuming that if a subscription is held for less than one year, then it is not profitable. Making an assumption that is not stated is the strategy used in this argument.

> teenagers cancel their subscription within one year.
> subscriptions held less than one year are not profitable (assumption)
> thus, it is not profitable to sell magazines to teenagers.

B - incorrect - all the facts necessary to reach the conclusion are not presented. Just from the fact that teenagers cancel their subscriptions within one year, we cannot conclude that it is not profitable to sell magazines to them.

| 2 | REASONING STRATEGY |

A survey of five hundred men with pot bellies..

The argument employs which one of the following reasoning techniques?
A) It incorrectly applies its conclusion to all the men surveyed, a fact that is true only for ninety percent of them.
B) It uses facts about pot bellied men that were not surveyed to justify its conclusion.

ANSWER

Answer: A

A – correct
 survey done on five hundred men with pot bellies.
 ninety percent of them drink at least one carbonated drink.
 carbonated drinks cause stomach to expand.
 so, carbonated drink is the reason those surveyed have pot bellies.

Note carefully that ninety percent of men that were surveyed drink at least one carbonated drink. The remaining ten percent do not drink carbonated drinks. But, the conclusion is applied to all the surveyed men. Since a majority of them drink carbonated drinks, this argument assumes that it is the reason why everyone surveyed has a pot belly. This is an invalid conclusion.

B – incorrect – facts about pot bellied men that were not surveyed are not even mentioned.

Answers
© Gift Of Logic, Inc * Copying prohibited

3 REASONING STRATEGY

Sonia: Traveling is fun, exciting and adventurous..
Sean: You can learn about world cultures..

Which one of the following describes the difference between Sonia's and Sean's arguments?
A) They reach different conclusions using different premises.
B) They reach the same conclusion by citing different premises.
C) They reach their conclusions by using different assumptions.

ANSWER

Answer: A

A – correct – Sonia concludes that traveling is the best way to learn about culture. Sean concludes that using the Encyclopedia software is the best way to learn about culture. Both reach different conclusions using different premises.

B – incorrect – their conclusions are different.

C – incorrect – no assumptions are made in either argument.

4 REASONING STRATEGY

Recent reports indicate that the traffic accidents..

Which one of the following reasoning techniques was used by the argument presented above?
A) It makes an inference by assuming an accurate causal relationship.
B) It makes an inference by assuming an incorrect causal relationship.

ANSWER

Answer: B

traffic accidents in the city have increased by twenty five percent
there has been a surge of teenage drivers lately
therefore, the teenage drivers c→ increase in accidents (causal relation)

A – incorrect – the causal relationship that is inferred from the facts is incorrect. There is no information presented to link the teenage drivers to the increase in accidents.

B – correct - there is a surge in accidents and a surge in teenage drivers. But, there is no evidence to prove that the teenagers are responsible for the accidents. But, this causal relationship is incorrectly assumed to make the conclusion that teenage drivers are responsible for the increase in accidents.

5 REASONING STRATEGY

When vending machines were introduced..

Which one of the following is true about the reasoning in this argument?
A) It can be logically inferred from the premises.
B) It is not supported by the premises.

ANSWER

Answer: B

A – incorrect – note the word "also" in the sentence ".. these days they also dispense ..". This means that vending machines continue to dispense snacks in addition to lunch items. There is no premise that indicates that no one is buying snacks anymore that would justify the conclusion. The conclusion cannot be logically inferred from the premises.

B – correct – the reasoning is not supported by the premises. The premises do not give evidence to support the conclusion that vending machines should stop dispensing snacks.

Answers
© Gift Of Logic, Inc * Copying prohibited

6 REASONING STRATEGY

Tom: Mixing the two chemicals..
Jerry: I agree that there..

Which one of the following describes how Jerry's argument is structured with respect to Tom's argument?
A) It agrees with Tom's conclusion, but not with the premise.
B) It agrees with Tom's premise, but not with the conclusion.
C) Its conclusion is identical to that of Tom's.

ANSWER

Answer: B

Tom's conclusion is: The experiment is safe.
Jerry's conclusion is: The experiment is dangerous.

A – incorrect – it disagrees with Tom's conclusion.

B – correct – both of them agree that there won't be a major explosion, but they differ in their conclusions.

C – incorrect – the conclusions are different.

Answers
© Gift Of Logic, Inc * Copying prohibited

7 REASONING STRATEGY

Trains and airplanes ..

The argument employs which one of the following reasoning techniques?
A) It gives all points of comparison the same weightage.
B) It compares two groups based on items that are biased toward one group.

ANSWER

Answer: A

A – correct - On two factors, safety and price, trains win. On convenience, both are the same. On time factor, airplanes win. So, based on the total number of points in its favor, trains win. The same weightage is given to all points of comparison.

B – incorrect – the points of comparison, namely, time, price, safety, and convenience are not biased toward any one group, either trains or airplanes.

8 REASONING STRATEGY

Most of the dramas in which Sanchez acted..

Which one of the following most accurately describes the role played in the argument by the fact that a few of his dramas that he acted in recently did not do well?

A) It is a counter premise that helps to prove that he does not know how to act.
B) It is a counter premise that helps to prove that he is not a popular actor now.

ANSWER

Answer: B

Note that the "but" in third sentence is a counter premise that changes the argument's line of reasoning. The conclusion is that he is not a popular actor now.

A – incorrect – The counter premise does not prove that he does not know how to act. The conclusion that emerges after the counter premise is presented only mentions that he is not a popular actor now. It does not say anything about his acting skills.

B – correct – The conclusion that emerges after the counter premise is presented clearly mentions that he is not a popular actor now. Thus, the counter premise helps to prove the conclusion.

9 REASONING STRATEGY

Sam: No animal must be kept in a Zoo. Taking them away..
Walton: It is quite true that animals have feelings..

Which one of the following accurately describes Walton's response to Sam's argument that no animal must be kept in a Zoo?

A) It uses a counter premise to present new facts that help reach its conclusion.
B) It agrees with the premise and conclusion in Sam's argument.

ANSWER

Answer: A

A – correct – Walton agrees that animals have feelings and should be left to live in their natural habitats. He then uses the counter premise keyword "nevertheless" to present facts about the usefulness of having a zoo to raise money and concludes that it is ok to have a few animals in a zoo.

B – incorrect – Walton agrees with Sam's premise that animals have feelings and that they must be left to live in their natural habitat. But, Walton does not agree with Sam's conclusion that no animal must be kept in a Zoo.

Answers

10 REASONING STRATEGY

Principal: Teachers feel that students must not be burdened..

The reasoning in the argument
A) takes sides with one party's interests while disregarding the other.
B) seeks a solution that would resolve the difference between the two parties.

ANSWER

Answer: B

A – incorrect – the argument does not take sides with either the teachers or the parents. Instead, it states that a solution that works for both the teachers and students must be found.

B – correct – this choice seeks a solution that would resolve the difference between the teachers and parents. This is what the principal says in her conclusion. Note that the reasoning of an argument is expressed in its conclusion.

11 REASONING STRATEGY

Nowadays, mobile phones come..

The fact that cameras in mobile phones have been misused plays which one of the following roles in the argument?
A) It is a premise provided to support the conclusion that cameras found in mobile phones must be banned.
B) It is a premise that supports the conclusion that it is illegal to use mobile phone cameras.
C) It is a counter premise provided to justify the conclusion.

ANSWER

Answer: C

A – incorrect – that the cameras found in mobile phones must be banned is not the conclusion.

B – incorrect – the conclusion is not that it is illegal to use mobile phones that have a camera in them. The conclusion only seeks new laws to define the legality of taking pictures with these cameras.

C – correct – note the use of the words "in spite of" in " in spite of this, these cameras have been misused …". This is a counter premise, that is presented to justify the conclusion that new laws are needed to define the legality of using the cameras in mobile phones.

Answers
© Gift Of Logic, Inc * Copying prohibited

12 REASONING STRATEGY

Alexia: Magicians are great ..
Alex: Magicians are cheats..

Alex counters Alexia's argument by
A) discrediting magicians as cheats because they use hypnotism.
B) claiming that magic and hypnotism are the same.

ANSWER

Answer: A

A – correct – Alexia claims that magicians are great. Alex counters this by saying that magicians are cheats because they use hypnotism.

B – incorrect – in fact Alex argues that magicians cheat by using hypnotism, thereby implying that magic and hypnotism are not the same.

13 REASONING STRATEGY

Parent: Toy manufacturers must provide..

The statement that suggests that toy manufacturers must provide computer games plays which one of the following roles in the argument?
A) It is the conclusion of the argument.
B) It is a sub conclusion of the argument.
C) It is a premise of the argument.

ANSWER

Answer: A

A – correct – the conclusion of the argument is the first statement. Although there are no conclusion indicators such as "Therefore", "Thus" etc, this is the conclusion as it naturally follows from the premises. You can arrange the premises and notice the conclusion. Sometimes, conclusion is stated first and then premises are presented.

> plastic and wooden toys come with lots of pieces
> they provide some fun initially
> they get strewn around the house quickly
> they are very small and can be swallowed by kids
> therefore, manufacturers must provide computer games as an alternative

B – incorrect – there is no other conclusion in the argument.

C – incorrect – this is not the premise of the argument.

Answers
© Gift Of Logic, Inc * Copying prohibited

14 REASONING STRATEGY

Mechanic: The chain that connects the two gears..
Supervisor: That cannot be true. As you can see, some..

The supervisor disagrees with the mechanic by
A) making a claim that cannot be inferred from the premises.
B) casting doubt on the mechanic's qualifications.
C) citing a different cause than the one cited by the mechanic.

ANSWER

Answer: C

A – incorrect – the supervisor makes a claim (conclusion) that can be inferred from the premises. He says that there are cracks in some of the links and that the chain snapped because of these cracks.

B – incorrect – the supervisor does not question the qualifications of the mechanic.

C – correct – see the causal diagram below. They cite different causes for the same effect (the snapping of the chain).

Mechanic: chain made of copper (weak material) c→ chain snap
Supervisor: cracks in chain c→ chain to snap

The symbol c→ represents the cause and the effect. Refer to the book titled "Critical thinking & Logical reasoning Primer" for a discussion on causal reasoning (cause and effect).

Answers
© Gift Of Logic, Inc * Copying prohibited

1 SUDOKU

Solve the following Sudoku. A correctly solved Sudoku has numbers 1-9 appearing only once in each row, each column and each 3x3 grid. Solving Sudokus will help you to gain valuable analytic skills.

5	3	4	6	8	7	1	2	9
6	9	2	5	3	1	4	7	8
8	7	1	4	9	2	6	3	5
2	4	7	8	1	6	9	5	3
9	6	5	7	2	3	8	4	1
1	8	3	9	4	5	2	6	7
7	5	8	2	6	9	3	1	4
3	2	9	1	5	4	7	8	6
4	1	6	3	7	8	5	9	2

Answers

© Gift Of Logic, Inc * Copying prohibited

2   SUDOKU

Solve the following Sudoku. A correctly solved Sudoku has numbers 1-9 appearing only once in each row, each column and each 3x3 grid. Solving Sudokus will help you to gain valuable analytic skills.

9	4	7	6	3	5	1	8	2
1	6	3	2	8	9	5	4	7
8	5	2	7	1	4	6	3	9
5	2	9	8	6	1	4	7	3
3	1	8	9	4	7	2	6	5
6	7	4	3	5	2	8	9	1
7	8	6	5	2	3	9	1	4
2	3	1	4	9	6	7	5	8
4	9	5	1	7	8	3	2	6

Answers
© Gift Of Logic, Inc * Copying prohibited

3 SUDOKU

Solve the following Sudoku. A correctly solved Sudoku has numbers 1-9 appearing only once in each row, each column and each 3x3 grid. Solving Sudokus will help you to gain valuable analytic skills.

7	3	9	2	1	6	5	8	4
2	1	4	7	5	8	9	3	6
5	8	6	4	9	3	2	7	1
6	5	1	3	8	2	4	9	7
9	7	8	1	4	5	6	2	3
4	2	3	6	7	9	8	1	5
1	6	2	9	3	4	7	5	8
8	9	7	5	6	1	3	4	2
3	4	5	8	2	7	1	6	9

Answers
© Gift Of Logic, Inc * Copying prohibited

4 SUDOKU

Solve the following Sudoku. A correctly solved Sudoku has numbers 1-9 appearing only once in each row, each column and each 3x3 grid. Solving Sudokus will help you to gain valuable analytic skills.

3	8	9	1	7	6	4	2	5
2	4	7	5	8	9	3	1	6
5	1	6	4	3	2	7	8	9
1	2	3	7	4	5	9	6	8
9	5	4	8	6	1	2	7	3
7	6	8	9	2	3	5	4	1
8	7	5	3	1	4	6	9	2
4	3	2	6	9	8	1	5	7
6	9	1	2	5	7	8	3	4

Answers
© Gift Of Logic, Inc * Copying prohibited

5 SUDOKU

Solve the following Sudoku. A correctly solved Sudoku has numbers 1-9 appearing only once in each row, each column and each 3x3 grid. Solving Sudokus will help you to gain valuable analytic skills.

9	5	2	7	8	3	6	4	1
8	3	6	1	4	5	7	9	2
4	7	1	9	2	6	3	8	5
5	2	3	6	9	4	1	7	8
7	8	4	3	1	2	9	5	6
1	6	9	8	5	7	2	3	4
2	1	8	4	3	9	5	6	7
3	4	7	5	6	1	8	2	9
6	9	5	2	7	8	4	1	3

Answers

© Gift Of Logic, Inc * Copying prohibited

1   POSITIONING

Andrea, Bobby, and Cindy.. >> A,B,C

If Andrea sings first, then Bobby can not sing next >> A1→ ~B2

A,B,C
A1→ ~B2

1	2	3	
A	B	C	choice A
C	B	A	choice B
A	C	B	choice C
B	A	C	choice D

Answer: Choices B,C,D

Choice A cannot be true. See the second row in the table. This positioning of the singers violates the rule A1→ ~B2.

Choice B can be true. See the third row in the table. This does not violate the rule A1→ ~B2, because A is not the first to sing.

Choice C can be true. See the fourth row in the table. B is not the second person to sing.

Choice D can be true. See the fifth row in the table. The rule is not violated.

Answers
© Gift Of Logic, Inc * Copying prohibited

2 POSITIONING

Chris, Raj, and Walid .. >> C,R,W

Walid cannot be the last one to perform >> ~W3

Note carefully that the question asks to find out which of the choices <u>cannot</u> be true. So, the choices that are not possible are the correct answers.

Answer: B, C. Only these two choices cannot be true.

Diagram the scenario as follows. Each choice is shown in one row. Clearly, choices B and C have W in the third position. These choices violate the ~W3 rule and hence cannot be true. So, these choices are the correct answers.

C,R,W
~W3

	1	2	3	
	W	R	C	choice A
	R	C	W	choice B
	C	R	W	choice C
	W	C	R	choice D

Answers
© Gift Of Logic, Inc * Copying prohibited

3 POSITIONING

Three clowns, Mumbo, Jumbo, and Fumbo ..>> M, J, F
Jumbo must always be above Mumbo >> this rule, where one is above the other can be represented as:

Note that there can be someone standing between Jumbo and Mumbo.

1) Which of the following can be true?
Answer: A. With Fumbo in position 2, Jumbo can be in position 3 and Mumbo in position 1.

M,J,F

3	J	
2	F	
1	M	J

Choice B cannot be true as can be seen from the diagram. If J is in position 1 then M cannot be placed below J.

2) Which of the following cannot be true?
Answer: C. Diagramming the choices leads us to the answer quickly. Mumbo cannot be in step# 3, because it will violate the rule.

M,J,F

3	J	J	M
2	F	M	
1	M	F	

Answers
© Gift Of Logic, Inc * Copying prohibited

4 POSITIONING

Andrea, Brianna, and Cindy..

The rule "They must sit in the order of their heights", means that the tallest can sit in the left most position or in the right most position. This must be carefully understood as this is logically possible. The rule does not say that they must sit in the order of their heights from left to right or right to left.

Andrea is taller than Brianna. This can be represented as A>B where > means "taller". Brianna is taller than Cindy. This can be represented as B>C. These two facts can be logically chained together and represented as A>B>C.

1) Given that A>B>C and that they must sit in the order of their heights, the following are the possible seatings. This answers Question# 1.

1	2	3
A	B	C
C	B	A

2) Of the possible ways that the three can be seated, which of the following is always true?
    Answer: A) Brianna sits between Andrea and Cindy.

Choice A - Brianna sits between Andrea and Cindy, is true in both cases. The word "always" refers to the two possibilities shown above. Choice B - Cindy sits in the right most seat, and choice C - Andrea sits in the left most seat are both not always true. Cindy and Andrea can sit in either the left most or the right most seats.

Answers

5 POSITIONING

Vivek, Brendon, and Farhad.. >> V, B, F

1) If Vivek must play before Farhad and Brendan must play after Farhad, then which of the following orders of performance is possible?

Answer: A
The rules in this question can be represented as follows:
Vivek must play before Farhad >> V-F
Brendan must play after Farhad >> F-B
We can chain the two rules logically as V-F-B. Note that V-F means that one or more performers can perform between V and F. Keeping the chained rule V-F-B in mind, we can answer the question.

   A) Vivek, Farhad, Brendan, John >> possible - Answer
   B) Vivek, John, Brendan, James, Farhad >> violates rule V-F-B
   C) Farhad, Brendan, John, Vivek >> violates rule V-F-B

2) In the festival, if Vivek plays before Farhad and Brendan plays after Vivek, then which of the following is possible?
Answer: B,C
The rules can be represented as V-F and V-B. Note that this cannot be chained as V-F-B since we do not know if F would play before or after B. We can only apply the two rules, V-F and V-B separately to the following choices.

   A) Brendan, Vivek, Farhad >> violates V-B rule
   B) Vivek, Brendan, Farhad >> possible - Answer
   C) Vivek, Farhad, Brendan >> possible - Answer

6 POSITIONING

Rudy, Puppy, Tommy, and Tony..>> R,P,To,Ty
Rudy must always sit in the third spot>> R3
Tommy and Tony must always sit in consecutive positions >> ToTy ∦ TyTo

1) In which of the following positions can Puppy be seated?
Answer: D) fourth. See the diagram below.

R,P,To,Ty
R3
ToTy ∦ TyTo

1	2	3	4
To	Ty	R	P

2) Which of the following is always true?
Answer: A) Puppy always sits in the fourth position. If Puppy sits in any other position, then To and Ty cannot be seated together.

R,P,To,Ty
R3
ToTy ∦ TyTo

1	2	3	4
Ty	To	R	P

3) If Rudy has to sit in the second spot instead of the third spot, then Puppy can sit in which of the following spots?
Answer: A) first.
The rules now are R2 and ToTy ∦ TyTo. To and Ty need 2 spots, 3 and 4. So, Puppy has to sit in the first spot only.

R,P,To,Ty
R2
ToTy ∦ TyTo

1	2	3	4
P	R	To	Ty

Answers
© Gift Of Logic, Inc * Copying prohibited

7 POSITIONING

Three mice get ready to sneak..

The rule "Either Rambo1 or Rambo2 must be in the middle" can be represented as R1@2 ǁ R2@2.

R1,R2,R3
R1@2 ǁ R2@2

1	2	3
R2ǁR3	R1	R2ǁR3
R3	R2	R1
R1	R3	R2

1) If Rambo1 is in the second position, what is the position of Rambo2?
Answer: Since R1 is @ 2, R2 can be at either 1 or 3. See the second row in the diagram. R2 can be in 1 or 3, but not both.

2) If Rambo3 is in the first position, then Rambo1 must be in the second position.
Answer: B) False. See the third row in the diagram. If R3 takes the first spot, then R1 can be at the second spot or the third spot.

3) Is the following order of the mice possible as they sneak into the kitchen?
            Rambo1, Rambo3, Rambo2

Answer: B) No. See the fourth row in the diagram. This violates the rule R1@2 ǁ R2@2.

Answers
© Gift Of Logic, Inc * Copying prohibited

8 POSITIONING

Three boxes labeled as Golf balls, Tennis balls, and Ping-pong balls... There should be at least one box below the boxes labeled "Tennis balls" and "Ping-pong balls". This rule can be represented as follows:

    T   P
    __   __

where __ means that there should be at least one box below T. In this case, since there are only two other boxes, there could be a minimum of one box and a maximum of two boxes.

1) What is the lowest position in which the box labeled 'Tennis balls' can be placed? Answer: B) 2. See diagram below. There should be at least one box below T. So, the lowest position for T is 2.

    3
    2 T
    1 __

2) What is the highest position in which the box labeled 'Tennis balls' can be placed? Answer: C) 3. This is clear from the diagram below.

    3 T
    2 P
    1 G

3) What is the lowest position in which the box labeled 'Ping-pong balls' can be placed? Answer: A) 2. There should be one box below at the first position.

4) What is the highest position in which the box labeled 'Golf balls' can be placed? Answer: A) 1 If it is placed in position 2 or 3, then one of the other boxes cannot be placed.

Answers
© Gift Of Logic, Inc * Copying prohibited

9 POSITIONING

George has the option to play soccer..

Fill the table with 'captain' on the days in which George will be the captain.

Week	Monday	Tuesday	Wednesday	Thursday	Friday
1	captain		captain		captain
2		captain		captain	
3	captain		captain		captain
4		captain		captain	

1) For at most how many days in the month can George be the captain of the team?
  Answer: B) 10. He can be the captain on the days that are indicated in the table. This adds up to 10 days.

2) For at most how many days in the second week is someone other than George the captain of the soccer team?
  Answer: B) 3
On the second week, George is captain only on Tuesdays and Thursdays. So, on the remaining three days, some one else is the captain of the team.

3) George captains more number of days during the first week than he does on the third week.
Answer: B) False. George captains the same number of days during the first and third weeks.

Answers

© Gift Of Logic, Inc * Copying prohibited

10 POSITIONING

The scores of three students in the Science, Math, and English..

Subjects	Student 1	Student 2	Student 3
Science	90	95	80
Math	80	70	90
English	70	60	60
Total -All	240	225	230
Total -Math and Science	170	165	170

1) Based on the total scores, the ranking of students is
    Answer: B)   Student 1, Student 3, Student 2

2) The ranking of students based on their math and science scores is
    Answer: C) Student 1 and Student 3, Student 2

Note that students 1 and 3 both have scored the same total in math and science combined. So, their ranking will be same.

3) If Student 2 had scored 76 in Math, then the rank of Student 1 based on Math and Science totals would be which one of the following?
    Answer: B) 2

If Student 2 had scored 76 in Math, then his total score in Math and Science would be 171 which would place him in the first rank. Student 1 and Student 3 both have a score of 170 in Math and Science, and so they will be in the second rank.

Answers

© Gift Of Logic, Inc * Copying prohibited

1 GROUPING

A sales manager sells his products..

The sales manager must visit at least one city. That means, in his trips, he can visit a minimum of one city and a maximum of three cities.

The rule "If he does not visit city P, then he must not visit city Q" can be represented as ~P→ ~Q. The contrapositive of this conditional statement is Q → P. That is, the rule can be rephrased as "If Q is visited, then P must be visited". It is easy to remember and apply this contrapositive rule instead of the original rule ~P → ~Q.

1) Which of the following cities can the sales manager visit?
Answer choices: B, C, D , E

A) Q and R	violates Q → P rule
B) P and R	possible Q → P does not mean P → Q
C) P and Q	possible
D) P, Q, and R	possible
E) P	possible
F) Q	violates Q → P rule, if Q is visited then P must be visited.

Answers
© Gift Of Logic, Inc * Copying prohibited

2   GROUPING

Danny went to a dentist to have his teeth..

1) Which of the following sets of teeth can be pulled out of Danny's mouth?

The rule "If tooth T1 is not pulled out, tooth T3 will not be pulled out" can be represented as $\sim T1 \rightarrow \sim T3$. The contrapositive of this condition is $T3 \rightarrow T1$. That is, if he removes T3, then he will remove T1.

Answer: A, C

A) T1, T2, T3	possible
B) T2, T3, T4	violates $T3 \rightarrow T1$ rule
C) T1, T2, T4	possible $T3 \rightarrow T1$ does not mean $T1 \rightarrow T3$. T1 can be pulled without pulling T3.

Answers
© Gift Of Logic, Inc * Copying prohibited

3 GROUPING

Seeds of five different vegetables..

Note that at least three veggies must be planted. So, the minimum is 3 and maximum is 5. The other rules can be represented as follows.

If veggie-1 and veggie-2 are planted, then veggie-4 must not be planted.
>> veg1 & veg2 → ~veg4

If veggie-2 and veggie-3 are planted, then veggie-5 must be planted.
>> veg2 & veg3 → veg5

If veggie-5 is not planted, then veggie-4 must not be planted as well.
~veg5 → ~veg4 (contrapositive is veg4 → veg5)

1) Which of the following groups of veggies can be planted in the garden.
Answer: A

A) veggie-2, veggie-3, veggie-5	can be planted
B) veggie-2, veggie-3, veggie-4	cannot be planted, violates veg2 & veg3 → veg5 rule
C) veggie-2, veggie-3, veggie-1, veggie-4	cannot be planted, violates veg1 & veg2 → ~veg4
D) veggie-1, veggie-2, veggie-3, veggie-4, veggie-5	cannot be planted, violates veg1 & veg2 → ~veg4

Answers

© Gift Of Logic, Inc * Copying prohibited

4	GROUPING

Andrew has a choice of eating..

Andrew can eat some, but not all the ice-creams. This means he can eat a minimum of 0 ice creams and a maximum of 2 ice creams.

He cannot eat IC-3 and IC-1 together. This can be represented as
~ IC-3 IC-1

1) Which of the following represent the different combinations of ice creams that Andrew can eat?
Answer: B, E
    A) IC-1, IC-3 >> violates rule ~ IC-3 IC-1
    B) IC-2, IC-3 >> can eat
    C) IC-3, IC-1, IC-2 >> cannot eat all.
    D) IC-3, IC-1 >> violates rule ~ IC3 IC1
    E) IC-2 >> can eat

5	GROUPING

There are five theme parks, P1, P2, P3, P4, and P5..
At most three of the parks can be visited.
At least one of P2 and P3 must be visited. >> P2 || P3

1) Which of the following represent the different theme parks that Paula can visit?
    A) P1, P4, P5 >> incorrect, either P2 or P3 must be visited
    B) P2, P4, P5 >> correct
    C) P4, P3, P2 >> correct, both P2 and P3 can be visited.

Answers
© Gift Of Logic, Inc * Copying prohibited

6 GROUPING

Movies m1, m2, and m3 are shown in the mornings..

Neil must watch at least three, but not all movies. This means he must watch a minimum of 3 movies and a maximum of 5 movies. >> min3, max5

In, addition, he must watch at least two movies in the morning and at least one in the afternoon >> 2M, 1A where M is Morning, A is Afternoon

Remember the rules min3, max5, 2M (at least), 1A (at least) and evaluate the choices.

1) Which of the following choices does Neil have to watch movies?
Answer: A, C

A) m1, m2, m3, m4	correct choice
B) m1, m4, m5	Incorrect - violates 2M rule
C) m1, m3, m4, m5	correct choice
D) m1, m2, m3	Incorrect - violates 1A rule
E) m1, m3	Incorrect - violates min3 rule

Answers
© Gift Of Logic, Inc * Copying prohibited

7	GROUPING

A minimum of five students..

If the minimum 5 students are selected, 20 students will not be selected.
If the maximum 10 students are selected, 15 students will not be selected.

1) What is the maximum number of students that will miss the selection?
     Answer: 20.
2) What is the minimum number of students what will miss the selection?
     Answer: 15.

8	GROUPING

A team comprising of a minimum of four students..

A and B must be picked together >> AB
B and E must be picked together >> BE . These two can be logically chained together as ABE. That is, A,B, and E must be picked together.
E and F must not be picked together >> ~EF

1) Which of the following set of students can be picked? Answer: B
    A) A, B, E, F >> violates ~EF
    B) A, B, C, E >> correct, A,B and E are picked. E and F are not.
    C) A, B, C, D >> violates ABE
    D) B, E, C, D >> violates ABE
    E) C, D, E   >> violates ABE
    F) B, C, F   >> violates ABE

Answers
© Gift Of Logic, Inc * Copying prohibited

9 GROUPING

Four people are selected from two groups..

The rules are represented as follows
- 1-ABCD, 2-EFGH
- 2 from each group
- A → GH
- D → EF

1) If A is selected from group-1, then which of the following can also be selected?

Answer: A) B.
If A is selected, GH is selected from group 2. This means that we need one more from group 1. Choice B is incorrect because if we select D then we will have to select EF, but we can select only two from each group and GH has already been selected. If we select B, the rules will be satisfied.

2) If D is selected from group-1 and B is not selected, then which of the following can also be selected?

Answer: B) C.
If D is selected, EF is selected. B is not selected thereby leaving us with A and C. If A is selected, GH will have to be selected, which is not possible as there should be only two from each group. If C is selected, then along with D, it will satisfy the rules. CD, EF will be correct composition of the two groups. So, choice B is the correct answer.

Answers
© Gift Of Logic, Inc * Copying prohibited

10 GROUPING

A, B, C, and D..

P	Q	R
A B C D	E F G H	I J K L

Four members must be selected from these three groups.
At least one member must represent each of the three groups.

1) If B, G and K are selected and there should not be two members selected from Group P, then the fourth member can be from which of the following sets?

Answer: A) (E,F,H) or (I,J,L). Since three members, B, G and K are already selected, we need one more. Rule says two members cannot be selected from P. So, this new member has to come from Q or R. Since members cannot be selected twice, we are left with E,F,H or I,J,L to pick the fourth member. Choice B-(A,C,D) is incorrect because we cannot select two members from group P. Choice C-(A,C,K) is incorrect because K has already been selected.

2) If B, I, and K are selected and there should not be two from either P or Q, then the fourth member can be from which of the following sets?

Answer: A
   A) Q >> correct, the fourth member can come from Q
   B) (A,C,D) >> incorrect, there cannot be two members from P
   C) (J, L) >> incorrect, at least one member must be selected from Q

Answers

© Gift Of Logic, Inc * Copying prohibited

11 GROUPING

L, M, N, and O are members of Group P.
R, S, T, U, and V are members of Group Q.

A team of four must be selected from groups P and Q.
There should be more members from group P than from group Q.
At least one from P and at least one from Q must be in the team.

```
   P            Q
  LMNO        RSTUV
```

1) How many members can be selected from P and Q?
Answer: A) three from P, one from Q. Since the other choices do not have more members of P than Q, they are not correct.

2) If L and R are selected into the team, then the remaining two will come from which of the following groups?

Answer: C) (ONM). Since there should be more members from P than from Q, for a team of 4, this means that there should be 3 from P and 1 from Q. Since R has already been selected from Q, we cannot select anymore from Q. The remaining two have to come from O, N and M. Choices A and B violate the rule that there should be more from P than from Q.

Answers

12	GROUPING

P	Q
L, M, N, and O	R, S, T, U, and V

1) If there should be as many members in the team from group P as there are from group Q, then which of the following represents a valid team?

Answer: B, D. Choice B-L,M,N,R,S,U is correct because there are three from each group. Choice D-L,T is also correct for the same reason. Choice A has more members from group P and choice C has more members from group Q.

13	GROUPING

P	Q
L, M, N, O	R, S, T, U

Y - 2P, 1Q
Z - 2Q, 1P

Sub-groups Y and Z have three members each.
In Y, there should be more members from P than from Q. This can be achieved by having 2 members from P and 1 member from Q. In Z, there should be more members from Q than from P. This can be achieved by having 2 members from Q and 1 member from P.

1) How many members are there in a group that is formed by combining the members of groups Y and Z?
Answer: A) - three from P, three from Q. Adding up the members in Y and Z gives us 3P, 3Q. Choice B - three from P, four from Q is incorrect.

Answers

PATTERN PERCEPTION

Question#	Answer
1	A
2	A
3	B
4	A
5	B
6	B
7	A
8	A

FIGURE FORMATION

Question#	Answer
1	A
2	A
3	B
4	B
5	A
6	B
7	A
8	B

Answers
© Gift Of Logic, Inc * Copying prohibited

PAPER FOLDING AND CUTTING

Question#	Answer
1	B
2	B
3	D

FIGURE MATRIX

Q#	Ans	Reasoning
1	A	after cooking, we get food; after drawing, we get a picture
2	A	several pages form a book; several flowers form a bunch
3	B	TV broadcasts picture ; radio broadcasts sound
4	A	one man from a crowd; one leaf from a tree
5	B	test tube, measuring cup, and burner are lab items; beaker is a lab item
6	A	heart, ribs, and bones are inside the body; brain is inside the body
7	B	balloons, cake, and caps are party items; gift is a party item
8	B	policeman, postman, and fireman are professionals; doctor is a professional.

RULE DETECTION

Question#	Answer
1	C
2	A
3	B
4	A

Answers
© Gift Of Logic, Inc * Copying prohibited

NOTES

NOTES

www.ingramcontent.com/pod-product-compliance
Lightning Source LLC
Chambersburg PA
CBHW080257180526
45167CB00006B/2569